BRINGING ARCHITECTURE TO THE NEXT LEVEL

Do your architectural designs foster well-being or discomfort in occupants? Do your designs innovate architecture technology in new ways? And do you make design decisions that pull from the latest scientific findings, including biomimicry design or where design meets neuroscience?

As an architect, your design process is vital to the quality and outcome of your built environments. Yet how do you know if you are creating optimal designs for well-being? Or strategic designs that uplift the future of architecture into new, more innovative realms?

In *Bringing Architecture to the Next Level*, you will discover how to solve the six greatest design challenges that hinder all architects. And when you unlock these six keys, you position your architectural design process to achieve its highest potential.

Within this one-hour guidebook, you will shift your mindset to reach breakthrough ideas, meet and predict occupant need using sensory design, leverage your design process to get more with less, and rethink technology to unleash your innovative edge. For both architecture and interior design professionals alike, *Bringing Architecture to the Next Level* will help you integrate innovation in architecture to create environments where people thrive.

Maria Lorena Lehman is a visionary author, designer, artist, and educator focusing on the links between environmental design, science, and emerging technologies. Lehman is the founder of MLL Design Lab and is author of the award-winning book, *Adaptive Sensory Environments*. Described as *"one of the leading experts on delivering exceptional occupant experience through smart building design,"* the work of Maria Lorena Lehman attracts an international audience as it bridges between theory and application to unlock what environments can do to benefit people in innovative ways. Lehman holds the degrees of Master in Design with Distinction from the Harvard University Graduate School of Design and Bachelor of Architecture, Cum Laude, from Virginia Tech. Lehman is recipient of the Harvard University Digital Design Prize for the *"most creative use of digital media in relation to the design profession."* Maria Lorena Lehman is internationally published and in numerous periodicals, including *The Architect's Journal, Esquisses Magazine, Architect Magazine,* and *Forbes.*

To learn more, visit **www.mlldesignlab.com**

FREE MASTERCLASS

Free MasterClass: Watch this masterclass by Maria Lorena Lehman that will forever change how you think about environments:

www.mlldesignlab.com/masterclass

BRINGING ARCHITECTURE TO THE NEXT LEVEL

BY
MARIA LORENA LEHMAN

LUMENSPIRE

Bringing Architecture to the Next Level
Third Edition (2020)

Copyright © Maria Lorena Lehman (2020).

First Edition published 2010. Second Edition published 2016.

ISBN: 978-1-7356006-0-4 (hbk)
ISBN: 978-1-7356006-1-1 (ebk)

Published by Lumenspire

Lumenspire is an imprint of MLL Design Lab, LLC
www.mlldesignlab.com

Cover and Interior Design: Copyright © Maria Lorena Lehman

For all architects who make this world an even better place.

How To Reach True Architectural Innovation

What you are about to read is the result of years of research and experience. With this book I intend to reveal what I think is of critical importance to help advance the architectural design discipline today.

We live in an era where there are many wonderful advancements taking place in different, and seemingly unrelated fields. I believe such advancements hold within them great opportunity once they are "bridged" strategically with key ingredients of architectural and urban design.

It is my goal with this book, *Bringing Architecture to the Next Level*, to introduce insights that open the eyes and expand the mind of architecture and urban design professionals, scholars, and design enthusiasts. Prepare to be energized with new workable design strategies that will lead you to true architecture and urban design innovation.

Here's to designing a better world,

Maria Lorena Lehman

Maria Lorena Lehman
Founder, MLL Design Lab, LLC

CHAPTER 1

Occupant-Centered Design

So much of your building's potential depends on your ability to meet your occupant's needs. As an architect your focus should always include your occupant in addition to designing for your building's budget, code regulations, developer's requirements, and client's goals. Through your design solutions you can become an important advocate for your building's occupant, incorporating their perspective as you design for their experience.

Your Own Design Process

As an architect today, you must solve complex problems in creative, resourceful, and aesthetic ways. With each project comes new design challenges, new construction methods, and new technologies, which all require continuous advancement in your way of thinking.

But how can you make your designs better? How can you improve from one project to the next? And how can your building designs go from simply

WHAT HINDERS YOU THE MOST?

Your Weakest Link.

When analyzing your own design process, it is often helpful to take a serious look at finding your biggest constraint.

You may have heard the adage: "You are only as strong as your weakest link." Well, this can be true.

By figuring out what is holding your designs back the most, you can make the effort to eliminate that constraint. In doing this, your overall design process and end-product will improve dramatically.

This stems from a business concept, but it holds true for design as well.

being good...to being **great**? To help us get to the bottom of this, I will simply start by posing the following question:

"What is your greatest challenge when designing architecture?"

This is also a question I recently asked to thousands of my *Journal of Design Insight* readers. Of course, there were a variety of responses; but there were some that I kept receiving over and over again. Hence, the following is my synopsis of the six primary challenges architects today told me they face when it comes to architectural design:

1. **Innovation:** Generating fresh ideas to create something new and useful

2. **Originality:** Avoiding unnecessary repetition while still having a collective body of work

3. **Awareness:** Removing bias so, as an architect, one's own experiences do not blur the correct answer

4. **Authenticity:** Filtering poor external influences and fads that can often find their way into one's designs

5. **Selflessness:** Taming one's ego by approaching each project with an open mind, ready to listen to other's ideas while realizing that architectural design is not simply about building a monument to oneself

6. **Communication:** Improving relationships with one's clients, which can impact the outcome of a project.

Within this book I will address these challenges by providing you with key concepts and mindset shifts so you can bring your architecture to the next level. The good news is that there is a key underlying factor that runs through each of the above challenges. It affects all of them, thus creating a domino effect.

By solving for this, you will see improvements rippling through your work. And it all begins by seeing your building designs through your building occupant's lens.

The Trust of Your Client

Clients play a critical part in your design process. After all, it is your relationship with them that can determine how seamlessly an architectural design project flows. Ideally, they should trust you and your expertise to carry out their needs successfully through your design creation.

The problem, however, arises when you as an architect neglect the actual people that will use your building from day to day. Sometimes your client is one of those people; other times they are not.

It is important to remember that just because you design for your client successfully, that does not automatically ensure that you did the same for your future building occupants.

As an architect, I invite you to keep both in mind as you design. That means really delving into the day-to-day functions and needs of your building occupants — how they perceive, think, feel, and behave.

If you do your job well, your designs will ultimately meet the needs of both your client and your day-to-day building occupants. Both will be happy, and your designs will be that much stronger.

Expanding Your Building's Potential

Buildings have an impact. Once they are built, they can transcend scale, time, and even culture. Although it sounds simple, it is important to remember that buildings are for people. They are for the people that inhabit them as much as they are for the people that pay for them. As a designer of built environments, it should be your mission to design to optimize occupant experience — helping to transform a so-called "basic building" into **great architecture**, no matter what the project size.

When you design from the occupant's perspective, your building will function better and take on a more profound beauty. Conversely, if you do not seriously consider your occupant's experience, your design will yield a simplistic environment — not truly healthy, aesthetically pleasing, safe, or very functional.

In the following diagram you can see why it is so important to incorporate your occupant's needs throughout your design process. You see, your building's potential depends upon your design process, and your design process' success depends upon your ability to meet your occupant's needs.

I know this concept may sound obvious, but when you look at how many buildings today

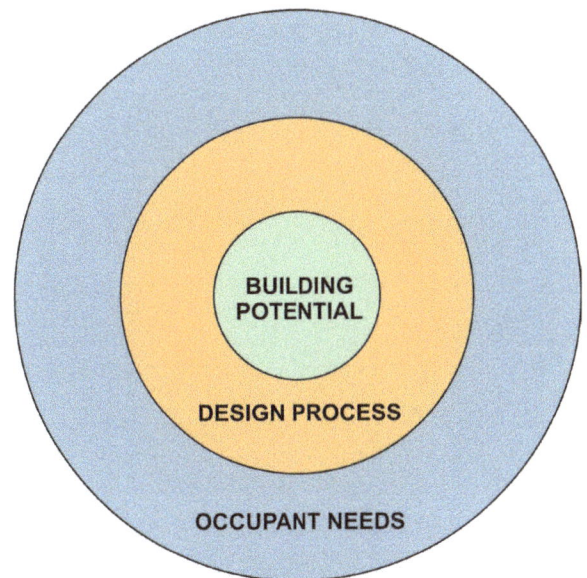

BUILDING POTENTIAL

DESIGN PROCESS

OCCUPANT NEEDS

Diagram: Drivers That Determine Building Potential

are "simplistic" and not "great," you begin to realize that occupant needs are not often prioritized as a primary concern.

When you strategically integrate your occupant's perception into your building's design, you will begin to notice how other factors determining your building's success are affected. The preceding diagram shows this type of domino effect where your occupant's needs constrain your design process, and your design process constrains your building's potential. In other words, the greater your ability to successfully and creatively meet your occupant's needs, the greater the chances are for your building to achieve its potential for success. This ultimately is a major factor in separating the "simplistic" from the "great."

The Significance of Your Occupant's Lens

As you design, you must consider so many requirements, meet so many different needs, and still make this all happen within budget and on time. Furthermore, your design must be "approved" through so many phases that your initial creative gesture can get stripped away during the design "process." So, you may ask yourself:

"*How can I design 'great' architecture that offers something 'new' with each project?"*

An interesting thing happens when you begin to consider your occupant's lens more strategically during your design process. Like a chain reaction of events, many of the factors that typically contribute to a "great" design, like aesthetic beauty, seamless functionality, economic strategy, and cultural presence, suddenly all fit into place.

TOWARD A WELL-ORCHESTRATED DESIGN

There are many simple instances where you can see that a designer did not follow through in fully understanding and incorporating their occupant's perspective. For instance, heating ventilation and air conditioning (HVAC) vents inappropriately located within a school's classroom can often emit interrupting noise that conflicts with teaching and disrupts other student learning activities.

While designing, these details might seem minor, but they all add up. As an architect, it is important to think of your occupant as not simply a user that carries out one or two tasks at a time. Try to think about how all of the details in your design come together as perceived by your occupant who inhabits your building day in and day out.

By deeply delving into what you can do for your occupant with your design, you will find that your design thinking will expand. You will begin to see more possibilities. And you will begin to generate better ideas that creatively solve problems and make occupants' lives better.

You will find that all six of the challenges architects face (innovation, originality, awareness, authenticity, selflessness, and communication) stem from your ability to understand and execute a design that enhances your occupant's experience. The more you know about how building occupants perceive and experience space, the more empowered you will be as an architect.

Your designs will naturally yield spaces that engage your occupants to think, feel, and behave where they want to and in the way they want to. Your designs will be more functional, timely, and beautiful.

One by one, you will be better able to tackle your biggest design challenges — you will be better able to innovate, avoid repetition, filter and eliminate internal bias, get rid of external preconceptions, be open to more possibilities, and even develop a better relationship with your client. In fact, your building's success can be proportional to the understanding you have regarding your building occupant's experience.

So, to keep your designs fresh, unique, and helpful to those who inhabit them, I invite you to continuously ask yourself:

"How can I learn more about how building occupants perceive my architectural designs?"

Balancing Architecture with Technology

It is important to remember that the definition of "occupant" encompasses a lot more than simply the people that **physically** inhabit your built works. In today's age, your "occupant" can mean anyone who "uses" your built environment — physically, virtually, individually, and/or collectively.

Your occupant's needs actually go beyond mere programming, and today there are a wide variety of methods and tools that you can use to transform your multi-dimensional occupant requirements into state-of-the-art, new, and innovative, architectural works.

New technologies are surfacing and changing the face of what architectural space can do for its occupants. Emerging innovations like nanotechnology with its smart materials or sensing ubiquitous computing technologies with their goal-oriented approach to new interactive and adaptive environments are two key examples of technologies that are impacting architectural design today.

Similarly, there are now new technological systems making information visualization, augmented reality, and virtual simulation possible. As is typical with such innovative technologies, they continue to refine themselves while also making their way into our field of architectural design.

Architecture and technology are most definitely linked — and have been for a very long time. Today we see technology positioned to affect almost every aspect of architecture, from the way we build with new materials to embedding sensors and actuators for interactive and adaptive spaces. Such technological advances are wonderful, but it is important to realize that architecture and technology do not *easily* integrate.

Today, there are many good buildings that do not reach "greatness" because technology was added on "after the fact," not fully understood and not unified with a building design's overarching ability to meet occupant needs. Unfortunately, when technology is not fully integrated into an architectural design, it not only looks like an afterthought, but it also functions like one too.

Additionally, the improper incorporation of technology can often feel invasive and intrusive for building occupants. This often results in "painful" architectural environments where a lack of true balance and integration between architecture and technology causes many problems, especially for unsuspecting building occupants.

Yet, new technology is not inherently bad. In fact, it can be quite a beneficial complement to architecture — an integral part of it.

"You must understand what technology does, what you want to make it do, and what effect that will have upon your occupants."

As an architect, technology can be your greatest ally or your greatest nemesis.

CHAPTER 2

Towards an Innovative Architecture

A problem arises when we begin to look at more specific aspects of the relationship between architectural design and emerging technology. As technology develops, it evolves differently as compared to architecture — and the two do not often integrate in complementary ways.

Is Technology Your Weakest Link?

A lack of design integration between architecture and technology will ultimately affect the overall functionality and beauty of an architectural space. Occupants will suffer as technologies get in the way of what a design is meant to accomplish. Can you imagine a hospital that does not foster healing? A school that does not foster learning? Or an office building that does not foster creativity or productivity?

Technology is frequently the weakest link in a building's design — which is unfortunate because it holds so much potential. Thus, by improving your implementation of architectural technology during your design process, your buildings will improve throughout, particularly

YOUR LEVERAGE POINT

It is in your design process that you have the most leverage to build a successful building.

I have often heard that "it takes a lot of money to build a poor building, but it doesn't take that much more to get it right."

This "little bit more" is all about leverage. And to maximize your leverage, you need to focus your design efforts on "occupant-centered" design.

Doing this will help to open your mind to new possibilities that will actually work — by honing in on the *right* problem at the *right* time.

A key catalyst for this is technology.

since emerging technologies are ubiquitously proliferating throughout environments, and are gaining even more transient abilities.

Four Major Challenges for You to Tackle

Why is it so difficult for some architects to fully take advantage of what architectural technology can do? First, technology is everywhere, yet, it has its own limitations. Second, many architects do not fully understand it, think it is important, or foresee its potential.

Without an educated approach to designing with architectural technology, your designs will likely be limited as you strive to incorporate base-functionality, but do not understand or foresee its full effects upon your building occupants.

There are, however, those of you who do strive to incorporate technology into your architectural designs in healthy and innovative ways. Still, there are many significant challenges for which you must solve:

- **Challenge One:** Designing for Different Lead Times

  Architecture takes a long time to build. Typically, it can take up to five years to construct just one building, while technology seems to be upgrading quickly — at faster and faster rates. Consequently, architecture and technology evolve in different ways, without much cross-communication during the design stages of each. Thus, technology is often "added" to architectural designs as an afterthought.

- **Challenge Two:** Building Construction versus Mass Production

  Technology is mass-produced, while architecture is usually unique. Technology seems to "spread" just about everywhere, while a successful architectural work is more difficult to reproduce or "share" globally. Consequently, architects get less feedback about how their designs actually work. Technology goes through many stages of developmental "user testing," which architecture does not. Hence, problems can arise when the two are set to work together in real time and in an actual space.

- **Challenge Three:** Keeping Up with New Technological Capabilities

  Architecture that does not integrate technology well remains predominantly inflexible — bringing with it a lack of varying "mobility" for occupant use. It is important for architects to understand the capabilities of emerging technologies because releasing "flexibility" within architecture helps a building work with factors like accessibility, sustainability, functionality, and aesthetics.

- **Challenge Four:** Negotiating Customization with Standardization

   Buildings are often "customized" as architects design for a client, occupant, community, culture, and so on. This personalization is embedded by the architect into the design solution itself. On the other hand, technology maintains a mass customization approach where certain "modules" can be personalized within the framework of an already designed system.

   Often, merging the two without compromising your design either aesthetically or functionally can be difficult.

As you can see, these four challenges have consequences that can affect your design and your process in both broad and specific terms. The key, however, is to know that they exist and to do your best to design consciously. That means staying informed and educated about architectural technologies as they develop, learning to think about fresh ways to incorporate such technologies to innovate your designs, and finding good ways to create, strategize, plan, test, and customize your building for your occupant.

Although technology may be your building's weakest link, it is important to remember that it is only one part of a complex equation that makes "great" design. However, another significant part to that equation is your occupant. In fact, not knowing enough about your future building occupant may be *your* weakest link as you design.

A deep understanding of how your occupants will perceive, think, feel, and behave with your spaces cannot be overemphasized. After all, when you are better able to hone your design talents to tap into technology's potential, you are better prepared to bring your architecture to the next level for your occupant.

"You must do more that simply keeping your occupant in mind while you design — you must understand what it is like to be them."

Stop Design Problems from Getting Past the Drawing Board

Can a lacking architectural design bring your occupants pain?

When architecture and technology do not fully fit, occupants can experience many "symptoms" as a result. A design "symptom" is defined as a negative effect yielded from the unsuccessful integration between architecture and technology (Lehman, 2017). Such negative effects are referred to as "symptoms" because they are typically perceived and felt by occupants. Unfortunately, they are indeed often a source of pain.

The following are three primary ways "symptoms" can develop when technology is not integrated properly with architecture:

- **The Sensory Deprivation Symptom**

  Most buildings that you see today do not truly consider all of an occupant's needs. Occupant experience is intrinsically linked with an occupant's senses, and often such senses are neglected during design. When occupant senses are neglected, your occupant's experience will deteriorate, and your design will not reach its full potential. The key is for you, as the designer, to understand how the senses work and how to apply this knowledge into your designs.

  You can think of a simple example of sensory deprivation by looking at how lighting technologies are often incorporated into some buildings. Often, natural and artificial lighting have not been well coordinated. For example, artificial lighting feels terrible for occupants as they must shop for items within fluorescent-lit retail stores, or suffer from not having enough task lighting within an otherwise flexible work space.

  Sensory deprivation can "break" a space. Thus, by deepening your understanding of how your occupant's sensory processes work, you will be better equipped to make the right design decisions that eliminate this symptom.

- **The Cognitive Overload Symptom**

  We are living in an age where information is growing at exponential rates (Lehman, 2017).

  With such an abundant amount of information and the emergence of virtual, interactive, and adaptive architectural spaces, I invite you as an architect to consider how your approach to incorporating such architectural advancements will affect your occupant.

  A simple example that illustrates why your *approach* can "make or break" your design is seen within office buildings — places where occupants are consistently experiencing cognitive overload.

  Workers are frequently bombarded by an array of incoming and outgoing information that they have to understand, filter, and react to. Instead of allowing technology to simply serve as a gateway for all this information, a forward-thinking architect can use interaction, visualization, orientation, or other techniques to help workers make "sense" of this information in real time — without always having to be tethered to a desk or other immobile office space anchors.

Architecture can take on a more proactive role to help occupants — particularly when it comes to cognitive overload.

- **The Overlapping Technology Symptom**

  The overlapping technology symptom is simple to understand but challenging to eliminate throughout the design of your building.

  Integrating multiple building technologies quite often means that meeting one demand will sacrifice another. A very simple example of this can be experienced through sound — specifically, through noise. In more complex building types, like a hospital, it becomes quite easy for technologies to "overlap," causing adverse experiential "side effects" for patients.

OVERLAPPING TECHNOLOGIES POLLUTE MANY HOSPITAL DESIGNS

To help you better visualize how overlapping technologies can be detrimental to architectural design, we will continue with the hospital example. Did you know that sleep is a critical factor that helps a patient to heal? Within hospitals, patients have a difficult time getting the sleep they need because noise is a major problem — doors open and close, hospital carts roll down the hallways, in-room medical technologies are continuously making sounds, televisions are on, and HVAC systems hum.

Some may say this is simply poor architectural design, but one would be surprised by how often this type of overlap with technology actually happens, even in those "high-end" design situations.

Finding Your Innovation Point

As technology continues to propagate and embed itself in a multitude of objects and environments around us, architecture will need to take the lead so technology does not blindly infiltrate environments, thus limiting the benefits of occupant-centered design.

Creating state-of-the-art environments is wonderful, if you do it well. In fact, with the right design approach, you can resolve many design problems with new technologies. The key is to shift your architectural technology mindset and approach while also deepening your knowledge about what it means to design for your occupant.

The good news is that we can uncover and develop many tools, findings, and methods available today to tackle many of these problems. By honing these, we as designers can bring architectural design to the next level.

"**W**ith technological growth, there should also be meaningful architectural growth. Fusing these with occupant-centered design is a healthy way to achieve **architectural innovation**."

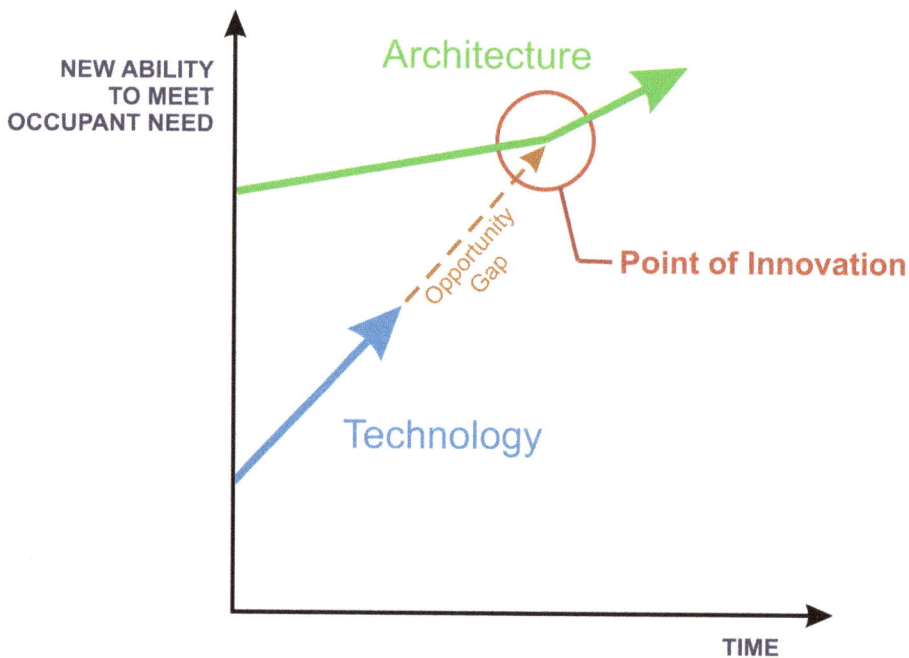

NEW ABILITY
TO MEET
OCCUPANT NEED

Architecture

Opportunity Gap

Point of Innovation

Technology

TIME

Diagram: Finding an Innovation Point

New applications for technology are constantly emerging in architecture. The question is whether or not architects know how to effectively design for them and then push that boundary.

CHAPTER 3

Master Design Communication

When one delves into understanding how occupant perspective and experience works, one's design process and outcome will reach far more innovative levels — and it is with innovation in mind that one can help architecture evolve by "communicating" through it.

Simplifying the Design Progress

Many truly great works of architecture push the envelope by redefining how technology works within them. Amazingly, there are also times when a design is so innovative, that new technologies must be created in order realize that architect's vision. Nevertheless, at their core, innovative designs improve upon the timely and timeless by helping to propel a culture forward. This is what I will refer to as **design progress**.

Architectural technology is an important part of design progress. As design paradigms shift over the years, architectural technology is intrinsically linked and changing as well.

Design visions guide architectural technology, while architectural technology informs design

HOW DESIGN VISION CAN GUIDE ARCHITECTURAL TECHNOLOGY

Did you know that silicone rubber caulking was invented specifically for weatherproofing Frank Lloyd Wright's Pyrex glass tubing in the Johnson Wax Research Tower?

The Dow Corning Company took three years in development and testing until they successfully decided upon the right material to use in 1955 (Lipman, 2003).

It still holds today, and now silicone rubber caulking is used widely for many purposes.

decisions. All three parts work together and play a major role in helping design progress take place.

What is Your Relationship with Architectural Technology?

As an architect, one should guide and learn from new architectural technology. Why?

When architecture and technology evolve together — in particular, focusing on when they are in a state of **design synergy** — a timely and timeless innovation point surfaces.

To help visualize this, architecture and technology are shown to communicate in the design synergy diagram below. Of course, they do not actually evolve at exactly the same time or in the same way. But it is interesting to simply explore what happens between the two when each informs the other.

As one can see, a positive feedback loop forms, thus helping to pull the best advantages from the other's key developments. To take architecture to the next level, it is important

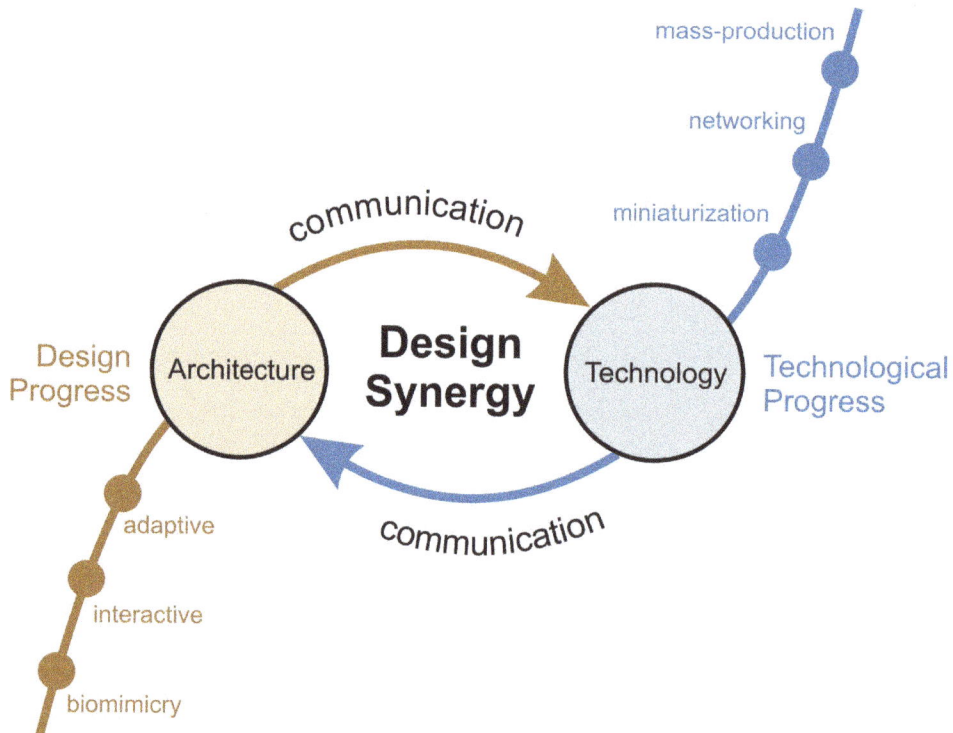

Diagram: Design Synergy Toward Architectural Innovation

for you to think about how you will both "feed" and "guide" technology through your designs.

Once you realize that it is you, the architect, who is responsible for bridging the gap between architecture and technology, then you will be in position to hone your skills so you can create moments where design synergy can flourish.

Again, this involves refining what I call your **architectural ingenuity**.

"**Y**our relationship with architectural technology will directly impact your ability to achieve greater innovation through your design work — in an **original**, **aware**, **authentic**, **selfless**, **clear**, and **helpful** way."

Boosting Your Architectural Ingenuity

Tackling such profound and meaningful principles as innovation, originality, awareness, authenticity, selflessness, and communication is fundamental to "great" design. All six of these challenges, experienced by so many architects, are difficult to conquer because they involve a continuous process — one which you will practice throughout your career.

The important thing to remember is that if you begin to get these right early on, you are well on your way to contributing meaningful architectural work for the world to experience. Simultaneously, you will also be building an accomplished architectural career in the process.

Yet, how can you work on all of these principles at once? The key is *communication*.

One must learn how to simply exchange information. Not just any information at any time, but the **right** information at the **right** time. When reading what I just wrote, this may look simple, but it is not so easy to do.

Take notice that communication is at the heart of the design synergy diagram. You see, it is communication that really facilities design progress, which, in turn, leads to innovation.

However, once you personalize this, and decide that you want to design more innovatively, it is communication that will help you to enter the innovation arena.

15

Chances are that you already communicate with your client. With your colleagues. With interns. With students. With history. With the future. With other cultures. With processes. With drawings. With analogies. With forms. With yourself.

Ultimately, your goal is to **communicate with your building occupants.**

Your Design as a Communication Portal

Architectural communication is both an inward and outward process by which you carefully listen and express. Communication involves most all aspects of architectural design. Your observations, questions, visions, explorations, guesses, experiments, and details all feed into your design instincts, with which you communicate.

There are also dimensions of architectural communication that are not quite as obvious. Have you ever thought of your design as a "communication portal"? It is through your built forms that you communicate with your building occupants — past, present, and future. While your previous designs inform the next, your communication skills set the foundation for your creative and visionary breakthroughs.

Communication is at the heart of all that you do, particularly when coupled with an "occupant-centered" design approach. By becoming an expert on how your architecture affects your occupants, you will be able to shift your building designs from being heavily biased toward the physical and visual into "moving" your occupants behaviorally, intellectually, emotionally, and spiritually.

AT THE HEART OF GREAT COMMUNICATION IS DECISION-MAKING

Your design tools are vital to helping you create innovative designs. It is with them that you communicate through many design channels, making a multitude of design decisions each day. Thus, it is vital for you to fine tune your ability to "think" and "focus" with them — on the right problems at the right times. Your goal is to unleash your ingenious ideas for maximum positive impact.

Learn to perfect this art and you will experience a ripple effect that travels through your design process. Your buildings will achieve greater and more profound levels of success. You will know where to look for insights, how to create a range of design solutions, and how to channel your expertise into selecting and developing the best ones. Your collective body of work will grow even more beautifully upon itself.

CHAPTER 4

Neuroscience: Your Building's Effect

By learning from other disciplines, you as an architect will become a more innovative and visionary thinker — particularly when delving into a field like neuroscience, which can teach you not only about scientific ways to be creative, but also about the inner working of the human mind and behavior.

Both are keys to deeply understanding how to meet your client's and occupant's needs, and then being able to design architectural forms that communicate with them at innovative, experiential levels.

A Cross-Disciplinary Approach

Just imagine if you could design buildings that foster your client's larger goals every time — like boosting productivity and creativity in the workplace, improving collaboration that leads to better learning in schools, or healing patients in less time and with higher quality of care in hospitals.

For these reasons, I invite you, as an architect, to more deeply reach across into other disciplines to gain a deeper understanding into what drives your building occupants. After all, they are the

THE HUMAN BRAIN

The human brain is one of the most complex designs on earth. What better design model to take cues from as you innovate?

From the air quality your building creates, to the aesthetic appeal it exudes, your building breathes just as the human brain thinks.

For your designs, the brain can teach you how to guide but not control, how to inspire but not to tire, and how to personalize but not forget the common good.

ones that use your buildings, and they are the ones who ultimately will determine the success of your built work.

For example, when you develop your innovative design talents, you increase the quality of the design so it works in multi-dimensional ways. When that happens, you have reached a state of design synergy within your built form — and often, that synergy involves an almost magical experience for your building occupants.

Design has the power to significantly uplift a person's well-being as it is processed through a person's experiential center, the human brain.

Why Neuroscience?

As an architect, I invite you to understand how the human brain works, how it affects your occupants, and how its extremely complex systems can inform your own architectural systems — especially as they become more intricate with the integration of new architectural technologies.

One of the most complex designs on earth is the human brain (Anon, 2014). It works seamlessly, and the design ideas that it holds within it continue to fascinate on a daily basis.

The human brain directly affects the work that you do as an architect, and the work that you do as an architect directly affects the human brain. As a designer, you will benefit from learning more about the neuroscience of human experience.

Bridging the "Gap" Between Architecture and Technology

Thinking about neuroscience as it relates to architecture actually makes your job as an architect much easier. New possibilities surface by allowing you to think of unique ways to bridge the "gap" between architecture and technology. Those four major challenges, described earlier in Chapter 2, now become quite doable.

Of course, many architects can simply merge architecture with technology to make a building work, but few can fuse the two together to harmoniously and synergistically create an architecture of profound beauty and meaning through functional and even transformational experience.

Achieving true design innovation is more than just constructing new forms that work. It is also about designing environments that are nurturing in ways that uplift the human spirit.

Thus, the latter is a key difference between architects that design for status quo function and the innovative few.

"Technology is not an excuse to sacrifice nurturing architectural design; instead it should contribute toward its advancement, and for this neuroscience is a bridge."

Technology can be designed as an advantage. And in mastering how to guide its development as it relates to built forms, the field of neuroscience can help the architectural discipline greatly.

In the diagram to the right, you can see how neuroscience supports a more nurturing and high-tech architectural design. At the point where architecture and technology meet with neuroscience, one's creative process is more likely to unleash design innovation that brings architecture to the next level.

A key goal to reaching this innovation point is to create buildings that embody and exude design synergy, where architecture, technology, and neuroscience fuse.

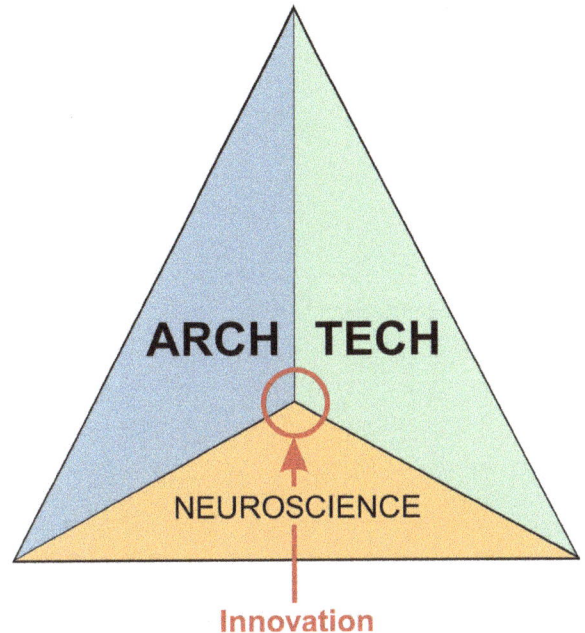

ARCH TECH

NEUROSCIENCE

Innovation

Diagram: Intersection for Reaching Design Synergy

What Does Neuroscience Reveal?

By learning about and applying neuroscience to your architectural designs, your building projects will grow in quality both functionally and aesthetically. You will find that your design process yields an architecture with more harmonized, complex, and flexible systems, as well as greater aesthetic sensibility and greater opportunity for meaningful occupant experience.

Neuroscience feeds into a multitude of dimensions that yield next-level architectural design, where neuroscientific findings provide insights that can help architects to more holistically integrate emerging technologies.

The reason neuroscience can offer so much to architecture is because it deals with a common theme that unites both architecture and technology — literally forming a "bridge" so each may fuse with the other. This commonality is what the brain is designed for. It is what helps us all to perceive the world around us.

It is all about transience and dynamics, or very simply put — *CHANGE*.

Adaptive Architecture for Tomorrow Can Help You Today

Architecture is gaining an increasing ability to change dynamically in real time, at many scales. The role of an architect is changing. Buildings and the processes that go into their creation are changing. As new technologies arrive, lead designers need to be well-equipped.

Designs will focus more on dynamic architectural behavior that contributes to an occupant's most meaningful experiences. And, as you can see in the diagram below, it is through the human senses that such architectural sensing technologies will work best.

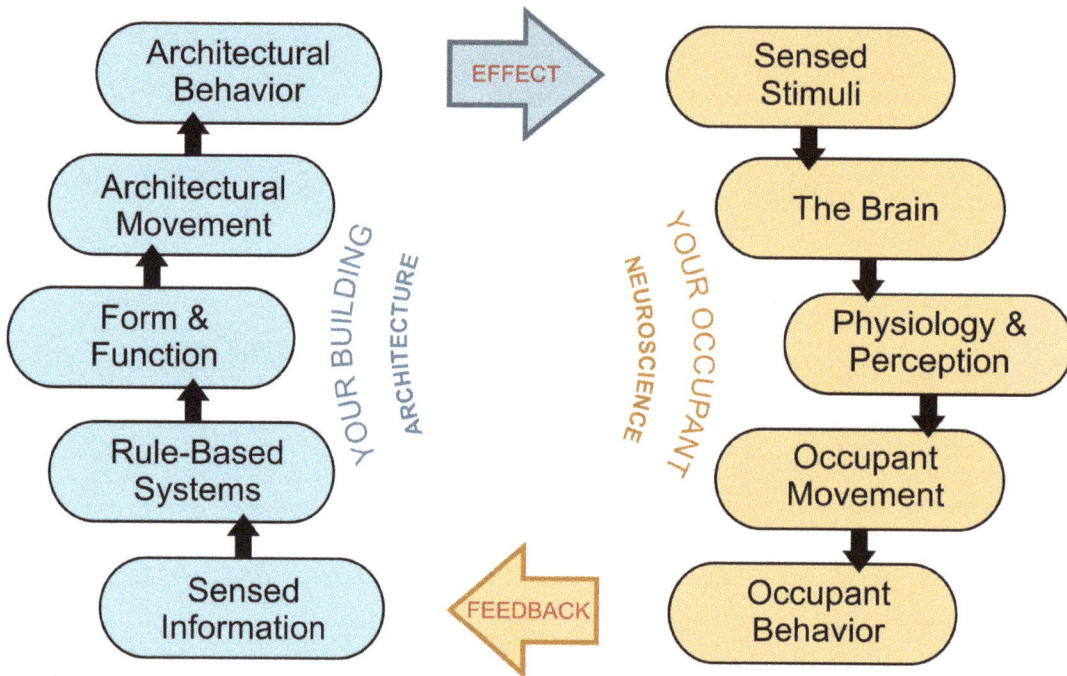

Diagram: The Behavioral Cycle Between Your Building and Your Occupant

As an architect, you design *change* — through expressed forms, useful functions, fused technologies, and sensed information. The architecture of tomorrow will call for an architect that can embed new kinds of rules and design behaviors together with design ingenuity. Neuroscience completes this picture, ultimately maximizing your design's potential by increasing your ability to improve life for your occupants.

CHAPTER 5

How to "Position" Your Architecture

Architect's roles are changing as architecture itself is morphing into spatial configurations that offer greater flexibility, personalization, and globalized mobility. But specifically, how can you enter into a heightened innovative design "state"?

Can Architectural Technology Increase Your Creativity?

Technological breakthroughs are changing the landscape of architectural design, and your role as "architect" is evolving as well. In order to not only keep up, but to also stay relevant and fresh, I invite you to envision what a future would be like for your occupants, both with and without such developments.

The following three questions become significant: Are you on the growing-edge of what all of these technological developments are capable of doing? Which ones will you integrate, and specifically for what architectural purposes? And how will you use them in your designs to improve life for your occupants?

Virtual reality, sensing devices, nanotechnology, ubiquitous computing, and new

> ### TECHNOLOGY = MASS CUSTOMIZATION
>
> New technologies that drive **Design on Demand** are important because they can give designers more control and flexibility over the small details that they previously could not specify due to budget, schedule, or fabrication constraints. By putting new mass customization methods on the forefront, **Design on Demand** is an example of how technology, when used wisely, can help architects design smarter (and faster) for their clients and occupants.

21

visualization technologies each bring a set of principles and capabilities just waiting for you, as a designer, to release their potential. Without good design integration, your architecture cannot flourish, and you may experience difficulty transcending what you "typically" do with features like light, material, acoustics, and even accessibility.

It is good practice for you to think "beyond of the box", as emerging technologies continue to surface. Although you may not utilize every innovative idea that you have, it builds your creativity to develop design "freedom" when it comes to integrating technology into architecture.

As you design by integrating all factors to create next-level architecture, it helps to think about what qualities new technologies can bring to your design palette that were not there before. Then, as you make design decisions, you can select which ones are most appropriate for your intended design challenge — and which ones can be "pushed further" to actualize your intended architectural vision.

The following are ten prime characteristics to "rethink" as you ask yourself how to innovatively integrate new architectural technologies into your design process.

INTERACTIVITY	Rethink "Response"	Think "Conversation"
TRANSIENCE	Rethink "Flexibility"	Think "Changing States"
KINETICS	Rethink "Motion"	Think "Behavior"
UBIQUITOUS	Rethink "Anchor"	Think "Seeds"
SELF-ACTUATING	Rethink "Maintenance"	Think "Prevention"
SCALABLE	Rethink "Material Properties"	Think "Nano-Scale"
CUSTOMIZABLE	Rethink "Individual Specification"	Think "Personal Preference" and "Stress Reduction"
NETWORKED	Rethink "Boundary"	Think "Rules"
VIRTUAL	Rethink "Simulation"	Think "Augmentation"
ADAPTABLE	Rethink "Responsive"	Think "Feedback"

Diagram: Ten Prime Characteristics to "Rethink" as You Integrate New Technology

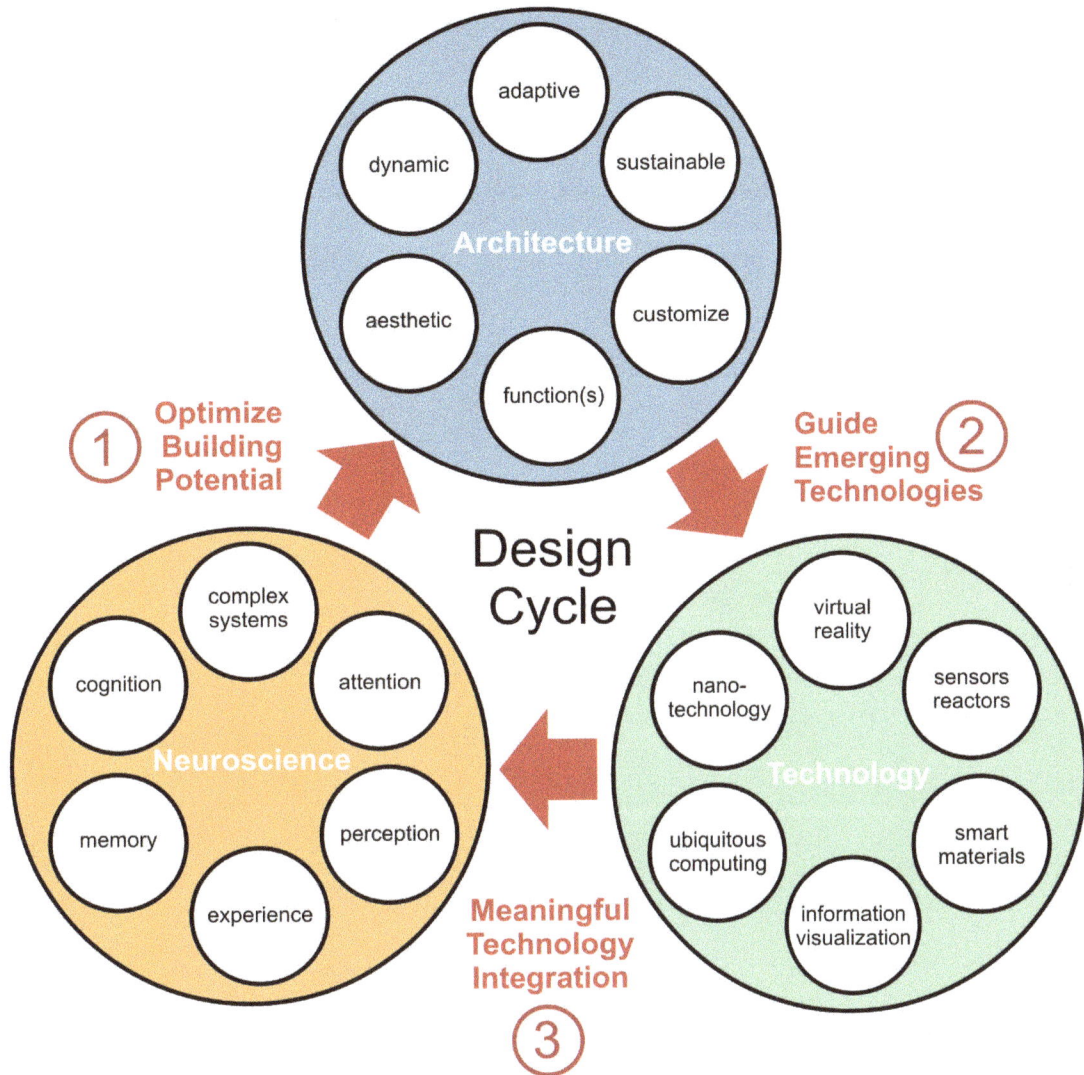

Diagram: How to Innovate Each Time You Make an Occupant-Centered Design Decision

Building Upon Your Previous Works

The key to all of this is the phenomenon of "sensing." Sensing involves communication, and communication involves behavior. With the latter in mind, as an architect, you become able to "optimize," "guide," and bring "meaning" for your occupant in unique and strategic ways. For instance, every time you make a critical design decision involving technology, your design will likely benefit from using neuroscience to inform its outcome.

As one can see in the design cycle diagram above, neuroscience helps architects to make more informed decisions for how to best uplift the well-being of an occupant. Note that the smaller grouped circles within the neuroscience, architecture, and technology headings of the diagram are simply design drivers chosen to illustrate this design cycle example.

It is important to bring one's own design vision and meaning into one's design cycle process. However, this leads to a fundamental next question: How can an architect ensure that their **architectural meaning** is both nurturing and innovative, while also building upon their previous works?

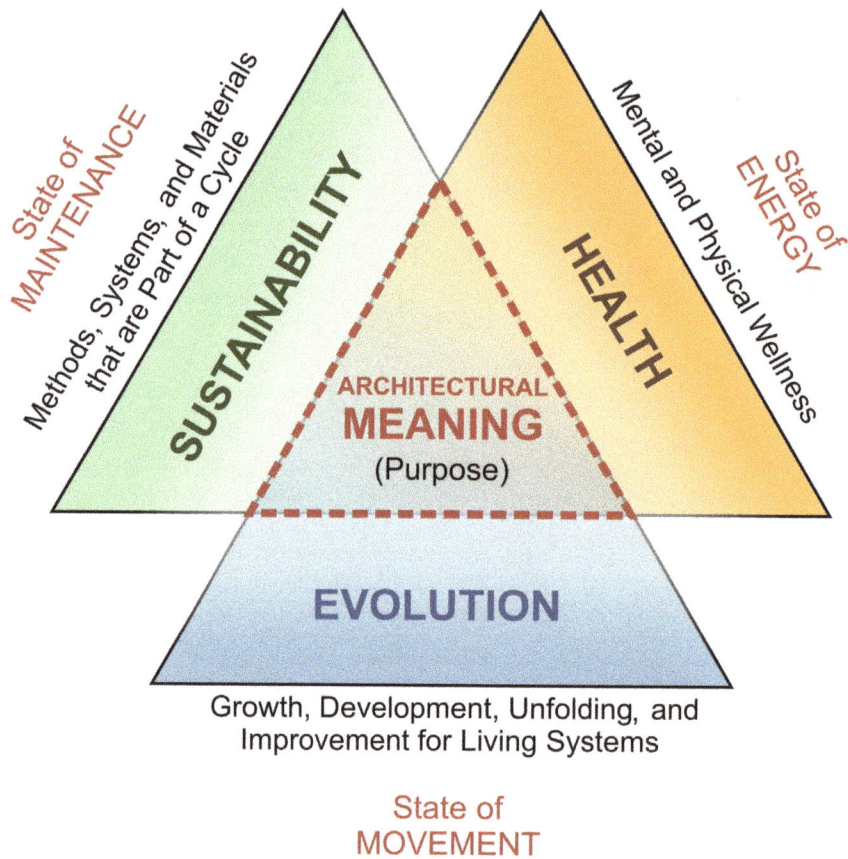

SUSTAINABILITY

State of MAINTENANCE

Methods, Systems, and Materials that are Part of a Cycle

HEALTH

Mental and Physical Wellness

State of ENERGY

ARCHITECTURAL **MEANING** (Purpose)

EVOLUTION

Growth, Development, Unfolding, and Improvement for Living Systems

State of MOVEMENT

Diagram: Three Primary Pillars for Occupant-Centered Design Innovation

Finding Architectural Meaning

"Architectural meaning" is a simple way to describe an otherwise profound and complex concept. However, for our purposes, "architectural meaning" refers to your design intention (or purpose) that drives your overall project's vision.

In the diagram on this page, you can see that your design's meaning (or purpose) can be a fusion between sustainability, health, and evolution. By incorporating this type of architectural meaning, your buildings will be both relevant to the present-time and forward-thinking.

As each architect embeds their own design talent, way of thinking, and unique perspective of today's world, this diagram serves as a starting point to guide thinking about where an architecture stands, both literally and figuratively.

It is important to understand that the three main areas shown in the diagram (sustainability, health, and evolution) each have sub-disciplines within them that contribute greatly to the architecture profession.

Architectural meaning, in this simple diagram, stands upon three "states," or conditions. Thus, to achieve your design's purpose, I invite you to incorporate each of these three states: maintenance, energy, and movement. By honing your design abilities to capture each of these architectural pillars, your buildings will best serve not only the needs of a building's occupants, but also more global planetary needs.

Building Purpose Rests upon These Three Foundational Pillars

The three states of maintenance, energy, and movement all support your architecture's purpose by tapping into "nature." After all, it is within nature that you learn about the states of maintenance, energy, and movement. As a designer, and particularly as an architect, nature serves as a resource that is both inspirational and vital to future innovation.

To reach next-level architectural innovation, I invite you to integrate nature at deeper levels into your design — to not only harmonize with it, but also to research its underlying design principles that make its life systems work.

Yes, there is science in nature. And technology is needed to apply its teachings to design.

Disciplines like biomimicry, nanotechnology, and neuroscience all encompass states of maintenance, energy, and movement. After all, each of these fields involve in-depth research targeting issues such as life-cycles, economy, and even aesthetics.

"A cross-disciplinary approach to finding your architectural meaning can often point you in new and more beneficial directions."

Biomimicry is a direct study of how to imitate nature to draw inspiration into current designs involving issues like self-organization, survival, and efficiency (Adams, 2017).

Biomimicry also encompasses neuroscience because the brain beautifully works with these issues every day.

Similarly, neuroscience, which is a direct study of how the brain works physiologically and perceptually, also couples well with the pillar of evolution because both involve growth, development, and decay every day.

The point is that it can help you, as an architect, to look into other seemingly unrelated disciplines to make new connections. As you gain versatility with the growing-edge findings and concepts of new technologies, biomimicry, and neuroscience, you will find that your paradigm will shift. You will design more effectively, and the collection of your architectural body of work will more beneficially impact the world.

Furthermore, as your architecture's purpose evolves from project to project, you will foster your own mindset shifts, which, in turn, spark more breakthrough ideas. For these reasons, it is important to keep nature in mind as you make design decisions. Learn to think of it in its fullest dimensions for it embodies and informs how you can grow as an architect, by improving your designs from project to project.

*"Nature gives us a real look into how processes can optimize, guide, and have meaning, and it is upon this that architecture, technology, and its design can advance and evolve for building occupants and our planet **by teaching us how to foster good health**."*

CHAPTER 6

Your New Role as "Sensemaker"

By taking steps to improve your design process, you will undoubtedly be taking action toward providing environments with *real value* for your clients and future building inhabitants. Ideally, your most valuable design solutions will stem from striving to foster good "health" — from helping your occupants to live healthier, to making sure your design solutions contribute to a healthier planet.

Why Buildings Today Fail and What You Can Do

In many ways, good design can yield good health. But good health can often decline with the emergence of bad design. For example, designing an office building that brings value to your occupants by promoting productivity can be done in two ways: the unhealthy way, or the healthy way.

In the unhealthy instance, an architect might primarily design for corporate cost savings with flexible spaces that can change as the employees' projects and teams evolve. However, in the healthier instance, a more innovative architect might also design for minimizing employee stress while also reinventing good socialization, communication, scheduling, and project management building design systems.

PROVIDING VALUE AS "SENSEMAKER"

As a forward-thinking architect in the twenty-first century, you will not only have to design for your occupant's senses, but you will also have to integrate sensing technologies into your building designs.

Thus, the "rules" by which you design will act as behavioral thumbprints that carry out your vision as "Sensemaker."

As you can see in these two examples, what you know about how employees perceive, experience, and behave in their workplace can be coordinated with what new technologies can do. The key is to know how to focus and integrate such knowledge and insight into your designs.

As a designer, I invite you to build a design process for yourself that emphasizes health as an undercurrent when making design decisions. Both your client and your occupant will thank you for your design's aesthetic, functional, **and healthy** innovative qualities that make your project valuable to their specific needs.

The key to designing for health is to fully consider how your design can care for your occupants more holistically, while also asking what your design can do that others are not doing.

The first step is to realize that your occupants have **many** dimensions that require architects to think beyond typical demands — like budget caps, code requirements, zoning laws, and review board approvals.

The Key to Your Occupant's Many Dimensions

Most buildings today predominantly bias one sense, in ways that do not leverage and at times even supress other senses (Pallasmaa, 2012). In these cases, this one sense is used to convey their building's "message."

*"Did you know that you have a better chance of providing real value to your occupants by considering **all of their senses** while you design?"*

Multi-sensory design will have a ripple effect throughout **all** of your work. Your buildings will be more functional, more beautiful, more cost effective, more sustainable, and more meaningful.

Additionally, you will create a stronger "connection" between your occupants and **your** way of designing because they will more fully engage when experiencing your environment.

For example, the sense of touch fuses with all of the other senses. When we touch something, we often say that we "feel" it. In reality, we feel with all of our senses, so there is actually "touch" involved however we navigate and experience architectural space.

If your occupants use all of their senses to "touch," then, your architecture should appeal to not only their visual sense, but to their aural (hearing), haptic (touch), olfactory (smell), and taste perception. In fact, these are *only five* from a multitude of senses experienced by the human body (Perry, 2018).

The sense of touch, alone, includes sensory modalities like temperature, vibration, and pain (Steward, 2000). Considering this in application, it would be helpful to you as a designer to understand why certain lighting types can cause "painful" experiences for your occupants. Think of patients in a hospital, children in a school, employees at work, or even shoppers in a retail store.

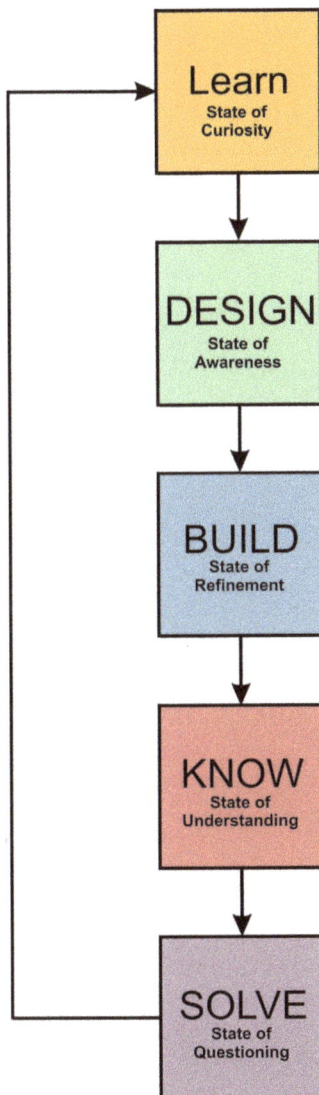

First, you must master *orchestrating* your designs, so they communicate helpful and healthy architectural stimuli to your occupants. The key is to learn how your occupants *physiologically* and *perceptually* respond to architectural space. In doing this, your occupants will flourish when engaging with your environmental design creations.

This is one reason neuroscience is such a helpful resource to tap into, not only to prevent your building from failing, but to propel it toward providing *real value* for the people and cultures it affects.

How to Drill into a Design Problem to Spark New Opportunity

When designing for occupants, you must constantly observe, and then ask "Why?" Why does a patient not heal in that hospital? Why do they not get enough sleep in their room? Why is there so much noise in the hallway? And why does their door keep opening and closing throughout the night?

Asking "why" will help you to become a better designer, by getting you to more subtly understand the nature of cause and effect. You will then see that you need to design for your occupant's challenges by learning about their root causes, because often a problem is really a symptom of some other root cause.

Diagram: Reinvent Your Process

By learning the techniques to properly "drill" into contextual issues that spark innovative solutions, one can impact the overall effectiveness of an entire design.

Thus, **context** is critical to figuring out why your occupants feel, think, or behave the way they do when within your space. Again, it is **neuroscience** that can lead you to truly understand your occupant's context — and how to design for it.

Neuroscience can help you figure out how to spot those critical design problems, turn them into design opportunities, and then orchestrate your own architectural design language — all so you can start *Bringing Architecture to the Next Level*.

With practice, you will innately be able to integrate most all of the principles, methods, and tactics explained in this book.

For instance, you will be able to determine your project's contextual core problems, elegantly communicate your own design meanings, fully integrate growing-edge architectural technologies, wisely interpret and design for behavioral response, and valuably "position" your work to create new and innovative design opportunities.

"The key is to reinvent the design aspects that you currently take for granted and to question the design aspects you use everyday."

BONUS RESOURCES

As the future unfolds, it is important to innovate new ways of *Bringing Architecture to the Next Level*.

For immediate access to even more educational resources that complement and expand this book, please visit MLL Design Lab at: **www.mlldesignlab.com**

Bibliography

Adams, D. (2017) *The best of biomimicry: Here's 7 brilliant examples of nature-inspired design.* [Online] Available from: https://www.digitaltrends.com/cool-tech/biomimicry-examples [Accessed 04 August 2020].

[Anon.] (2014) *The human brain is the most complex structure in the universe. Let's do all we can to unravel its mysteries.* [Online] Available from: https://www.independent.co.uk/voices/editorials/the-human-brain-is-the-most-complex-structure-in-the-universe-let-s-do-all-we-can-to-unravel-its-9233125.html [Accessed 04 August 2020].

Lehman, M.L. (2017) *Adaptive Sensory Environments.* Oxon/New York: Routledge: Taylor & Francis Group.

Lipman, J. (2003) *Frank Lloyd Wright and the Johnson Wax Building.* New York: Dover Publications, Inc.

Pallasmaa, J. (2012) *The Eyes of the Skin.* West Sussex: John Wiley & Sons, Ltd.

Perry, P. (2018) *Think you have only 5 senses? You've actually got about 14 to 20.* [Online] Available from: https://bigthink.com/philip-perry/think-you-have-only-5-senses-its-actually-a-lot-more-than-that [Accessed 04 August 2020].

Steward, O. (2000) *Functional Neuroscience.* New York: Springer Science + Business Media, LLC.

About the Author

Maria Lorena Lehman

Founder of MLL Design Lab

Maria Lorena Lehman is a visionary designer, author, artist, and educator focusing on the links between architectural design, science, and emerging technologies. Lehman is the founder of MLL Design Lab, LLC and is author of the award-winning book, *Adaptive Sensory Environments*. Maria Lorena Lehman is described as *"one of the leading experts on delivering exceptional occupant experience through smart building design."* Her research attracts an international audience as it bridges between theory and application to unlock what environments can do to benefit occupants in innovative ways.

Lehman holds the degrees of Master in Design with Distinction from the Harvard University Graduate School of Design and Bachelor of Architecture, Cum Laude, from Virginia Tech. Lehman is recipient of the Harvard University Digital Design Prize for the *"most creative use of digital media in relation to the design profession."* While at Harvard University, Lehman worked to innovate healing environments by researching at the nexus of architectural technology, digital media, and neuroscience within the Harvard Graduate School of Design, House_n, Media Lab at the Massachusetts Institute of Technology (MIT), and the Neuroscience Department of the Harvard Medical School.

Maria Lorena Lehman is internationally published and in numerous periodicals, including *The Architect's Journal, Esquisses Magazine, Architect Magazine*, and *Forbes*. Currently, her research looks for new ways environments can uplift quality of life by innovating experience through multi-sensory design, adaptive design, and emerging design process tools, strategies, and methods. At the heart of her work is a motivation to push the role of the environment into more proactive realms that empower people to thrive and achieve fulfillment at their highest potential. Maria Lorena Lehman has a vision for how interdisciplinary findings and innovations can converge together with creative and forward-thinking design processes to unlock more nurturing architectural futures.

To learn more, visit **www.mlldesignlab.com**

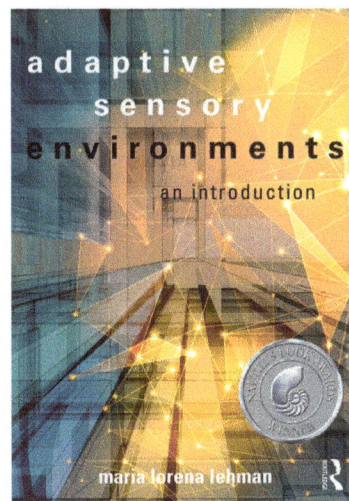

MARIA LORENA LEHMAN, AWARD-WINNING AUTHOR: *ADAPTIVE SENSORY ENVIRONMENTS*

FREE MASTERCLASS

Free MasterClass: Watch this masterclass by Maria Lorena Lehman that will forever change how you think about environments:

www.mlldesignlab.com/masterclass

www.ingramcontent.com/pod-product-compliance
Lightning Source LLC
Chambersburg PA
CBHW040256100426
42811CB00011B/1283